LIFTOFF! SPACE EXPLORATION

OUT OF THIS WORLD

MISSIONS AROUND THE SOLAR SYSTEM

BY ERIC KEPPELER

Gareth Stevens PUBLISHING

Please visit our website, www.garethstevens.com. For a free color catalog of all our high-quality books, call toll free 1-800-542-2595 or fax 1-877-542-2596.

Cataloging-in-Publication Data

Names: Keppeler, Eric.
Title: Out of this world missions around the solar system / Eric Keppeler.
Description: New York : Gareth Stevens Publishing, 2021. | Series: Liftoff! space exploration | Includes glossary and index.
Identifiers: ISBN 9781538258781 (pbk.) | ISBN 9781538258804 (library bound) | ISBN 9781538258798 (6 pack)
Subjects: LCSH:Astronautics–Juvenile literature. | Outer space–Exploration–Juvenile literature. | Space flight–Juvenile literature. | Space probes–Juvenile literature.
Classification: LCC TL793.K47 2021 | DDC 629.4–dc23

First Edition

Published in 2021 by
Gareth Stevens Publishing
111 East 14th Street, Suite 349
New York, NY 10003

Designer: Katelyn E. Reynolds
Editor: Abby Badach Doyle

Photo credits: Cover, p. 1 Withan Tor/Shutterstock.com; cover, pp. 1–32 (series background) David M. Schrader/Shutterstock.com; p. 5 VW Pics/Universal Images Group via Getty Images; pp. 7, 21 (inset) Universal History Archive/Getty Images; pp. 9, 15 (main) NASA/ullstein bild via Getty Images; p. 11 NASA/Interim Archives/Getty Images; p. 12 SSPL/Getty Images; p. 13 (main) AFP via Getty Images; p. 13 (inset) Keystone-France/Gamma-Keystone via Getty Images; p. 15 (inset) Space Frontiers/Getty Images; pp. 17 (main), 19 (inset) JSC/NASA; p. 17 (inset) AFP/AFP via Getty Images; pp. 19 (main), 23 MSFC/NASA; p. 21 (main) NASA/JPL; p. 25 (main) NASA/Johns Hopkins APL/Steve Gribben; p. 25 (inset) KSC/NASA; p. 27 NASA/JPL-Caltech/MSSS; p. 29 Joe Raedle/Getty Images.

Printed in the United States of America

CPSIA compliance information: Batch #CS20GS: For further information contact Gareth Stevens, New York, New York at 1-800-542-2595.

CONTENTS

WORDS IN THE GLOSSARY APPEAR IN **BOLD** TYPE THE FIRST TIME THEY ARE USED IN THE TEXT.

THE FINAL FRONTIER

For as long as people could look up at the tiny lights in the night sky, humans have wondered about the nature of outer space. At first, they didn't understand it at all. They thought it was just a ceiling over Earth, the way a roof covers a house. Some people thought that the stars were actually light from heaven shining through tiny holes in the sky.

Over time, people came to understand space is a great open area, and the little lights are stars and planets that are very far away. Our planet Earth and its solar system are in a **galaxy** called the Milky Way. This galaxy is so huge that it's very hard for scientists to count all of its stars.

INTERSTELLAR INFORMATION

SCIENTISTS THINK THE UNIVERSE IS 13.8 BILLION YEARS OLD, BASED ON THE AGE OF THE OLDEST OBJECTS IT CONTAINS AND ON HOW FAST IT'S EXPANDING.

THE MILKY WAY GOES BY OTHER NAMES IN DIFFERENT LANGUAGES. THE CHINESE CALL IT THE SILVER RIVER. IN SOUTHERN AFRICA, IT'S KNOWN AS THE BACKBONE OF NIGHT.

LOTS OF COMPANY

Most scientists believe there are at least 100 billion stars in our galaxy, although some believe there could be as many as 700 billion. The Milky Way is just one galaxy in our universe. It's hard to know for sure how many galaxies there are since space is so huge and scientists can only count the ones they can find. Most think there are between 100 billion and 200 billion galaxies in the universe.

EARLY YEARS OF STUDY

Almost 2,000 years ago, astronomers first started getting the idea that the flickering lights in the sky were actually stars and other planets. One of the earliest was an Egyptian scientist named Ptolemy, who was born in CE 90. He said Earth must be the center of the universe.

People believed that for centuries until Polish astronomer Nicolaus Copernicus suggested in the 1500s that the sun was the center of our solar system and Earth moved around the sun. Most people didn't believe him.

Italian astronomer Galileo Galilei was born in 1564. He developed the telescope to study other planets in our solar system, such as Jupiter and Saturn, and helped prove Copernicus's theory that Earth and other planets orbited the sun.

INTERSTELLAR INFORMATION

ENGLISH ASTRONOMER EDMOND HALLEY, BORN IN 1656, WAS THE FIRST TO REALIZE THE SAME COMET APPEARED OVER EARTH EVERY 75 YEARS. IT WAS NAMED HALLEY'S COMET AFTER HIM.

MOTION MAN

Johannes Kepler was a German scientist born in 1571. He discovered three laws of motion to describe the way planets moved around the sun. The first law says planets move in ellipses, or ovals. The second says a planet's orbit around the sun will always cover the same distance and amount of time. The third says the time a planet needs to orbit the sun depends on how far away it is.

ROCKET MEN

As time passed and technology improved, scientists realized how far away other planets were from Earth. They knew they would need something that traveled very fast if they wanted to visit other worlds, and so the rocket was invented.

One of the first people to understand the science of spaceflight was Russian mathematician Konstantin Tsiolkovsky in the early 1900s. Later, American Robert Goddard built liquid fuel rockets that were increasingly powerful. German scientist Hermann Oberth combined the math of space travel with the design of rockets.

In the 1930s, Wernher von Braun was a brilliant German scientist, but the ruling **Nazis** wanted him to make weapons instead. He later escaped to the United States and helped develop the Apollo rockets that went to the moon.

INTERSTELLAR INFORMATION

GODDARD'S FIRST ROCKET WAS LAUNCHED MARCH 16, 1926, IN MASSACHUSETTS. IT ROSE 41 FEET (12.5 M) AND TRAVELED A DISTANCE OF 184 FEET (56.1 M). HE LATER MOVED HIS WORK TO NEW MEXICO BECAUSE IT GOT TOO NOISY!

WHEN PROFESSOR ROBERT GODDARD FIRST ANNOUNCED THAT HIS ROCKET COULD TAKE OBJECTS TO THE MOON, NO ONE BELIEVED HIM. TODAY, HE'S WIDELY RESPECTED AS THE FATHER OF MODERN ROCKET SCIENCE.

SCIENCE FICTION OR SCIENCE FACT?

From the 17th to the 19th centuries, much exploration focused on our own planet—but outer space had captured the imaginations of several famous authors. This included Frenchman Jules Verne as well as Englishman H. G. Wells. These men used the knowledge that was available at the time to try to **accurately** describe what it would be like to travel through space to other worlds and meet alien life.

THE SPACE RACE

Interest in space took a huge jump on October 4, 1957, when the Soviet Union (modern-day Russia) launched the world's first artificial satellite, called Sputnik. It was only about the size of a beach ball and weighed 183.9 pounds (83.6 kg)—but it raced around our planet in 98 minutes.

The Soviet Union and the United States were the two largest superpowers in the world at the time. They often seemed on the **verge** of conflict—so when Sputnik was launched, Americans were afraid the Soviets could put **missiles** in orbit and use them to attack.

A month later, the Soviets launched Sputnik II, which carried a dog named Laika into space. The United States launched its first satellite, called Explorer I, the following January—and the race was on.

INTERSTELLAR INFORMATION

IN 1962, U.S. PRESIDENT JOHN F. KENNEDY CHALLENGED NASA TO PUT A MAN ON THE MOON AND RETURN HIM SAFELY HOME BEFORE THE END OF THE DECADE.

NASA'S FIRST ASTRONAUTS, CALLED THE ORIGINAL SEVEN, WERE, FROM LEFT: SCOTT CARPENTER, GORDON COOPER, JOHN GLENN, GUS GRISSOM, WALLY SCHIRRA, ALAN SHEPARD, AND DEKE SLAYTON.

THE BIRTH OF NASA

After the launch of Sputnik, the United States created NASA, the National **Aeronautics** and Space Administration. It started operating on October 1, 1958. Its purpose was to oversee space exploration. In the next few years, NASA launched many unmanned **probes** to explore other planets such as Venus and Mars. It also launched the first American astronauts. Alan Shepard became the first American in space in 1961 and John Glenn became the first American to orbit Earth a year later.

COSMONAUTS

In the early days of the Space Race, the Soviet Union was far ahead of the United States. It was the first to achieve many goals. Besides the Sputnik satellites, the Soviets also made the first probe (Luna 1) to fly by the moon in 1959. Luna 2 crashed on the moon's surface that same year.

Soviet astronauts were called cosmonauts. In 1961, Yuri Gagarin became the first man in space and the first to orbit Earth. Two years later, Valentina Tereshkova became the first woman in space while on a three-day mission aboard *Vostok 6.* On March 18, 1965, Alexei Leonov became the first person to go out for a spacewalk—although his suit ballooned and he had a hard time getting back into his ship!

INTERSTELLAR INFORMATION

THE SOVIETS LANDED A ROBOTICALLY CONTROLLED **ROVER**, CALLED LUNOKHOD 1, ON THE MOON IN 1970. FOR ALMOST A YEAR, IT TOOK PICTURES AND STUDIED ROCKS AND SOIL.

◀ VALENTINA TERESHKOVA

YURI GAGARIN, LEFT, WAS THE FIRST MAN IN SPACE, AND VALENTINA TERESHKOVA WAS THE FIRST WOMAN IN SPACE. GAGARIN DIED IN A PLANE CRASH IN 1968, WHILE TERESHKOVA BECAME A SUCCESSFUL SPACE ENGINEER.

SERGEI KOROLEV

THE MAN BEHIND THE SUCCESS

Sergei Korolev founded the Soviet space program. He oversaw many of these successful missions. Korolev was born in 1907 in Ukraine and studied rocketry at the University of Moscow. Soviet leader Joseph Stalin imprisoned Korolev and other scientists during World War II, but they still continued their successful work on aeronautics. When the war ended, Korolev was freed from prison and put in charge of designing Soviet rockets.

ONE SMALL STEP

The United States jumped ahead in the Space Race when it sent three astronauts—Frank Borman, Jim Lovell, and Bill Anders—on the *Apollo 8* mission to orbit the moon in December 1968.

Seven months later, the *Apollo 11* mission took Neil Armstrong, Edwin "Buzz" Aldrin, and Michael Collins to the moon. Armstrong and Aldrin got into the **lunar module** and flew it to the surface.

On July 20, 1969, the astronauts landed in an area called the Sea of Tranquility. Armstrong was first to walk on the moon, saying one of the most famous lines in human history: "That's one small step for a man, one giant leap for mankind." Aldrin followed him out 20 minutes later to become the second man on the moon.

INTERSTELLAR INFORMATION

APOLLO 17 IN 1972 WAS THE LAST MISSION TO THE MOON BEFORE THE PROGRAM RAN OUT OF MONEY TO BUILD ROCKETS.

IT TOOK THE *APOLLO 11* CREW 75 HOURS AND 56 MINUTES TO REACH THE MOON. ONCE THEY GOT THERE, MORE THAN 500 MILLION PEOPLE WATCHED THE MOON LANDING ON TELEVISION.

RISKY BUSINESS

Being an astronaut can be dangerous. On January 27, 1967, during a practice, the spacecraft caught fire and killed Apollo 1 astronauts Virgil Grissom, Roger Chaffee, and Edward White. Scientists learned from the accident to make new rockets safer. *Apollo 13* was launched April 11, 1970, carrying James Lovell, Fred Haise, and John Swigert. An explosion damaged their ship and kept them from landing on the moon, but the astronauts were able to return home safely.

EXPLORING OTHER WORLDS

In the early- to mid-20th century, the main focus was landing man on the moon—but scientists didn't forget about the other planets in our solar system. The first spacecraft to reach Venus was NASA's Mariner 2 in 1962. Venera 7 from the Soviet Union became the first vehicle to land on another planet when it set down on Venus in 1970.

Astronomers observed the large area of Olympus Mons on Mars for hundreds of years, but it wasn't identified as a volcano until 1971 by American probe Mariner 9.

NASA probes Voyager 1 and 2 were launched in 1977 and provided lots of photos and information about Jupiter, Saturn, Uranus, and Neptune and their moons. Both probes have now left the solar system, which means they're in **interstellar** space.

INTERSTELLAR INFORMATION

SKYLAB WAS THE FIRST AMERICAN SPACE STATION. MADE OF UNUSED APOLLO ROCKETS, IT ALLOWED ASTRONAUTS TO LIVE AND LEARN IN SPACE FOR UP TO 84 DAYS.

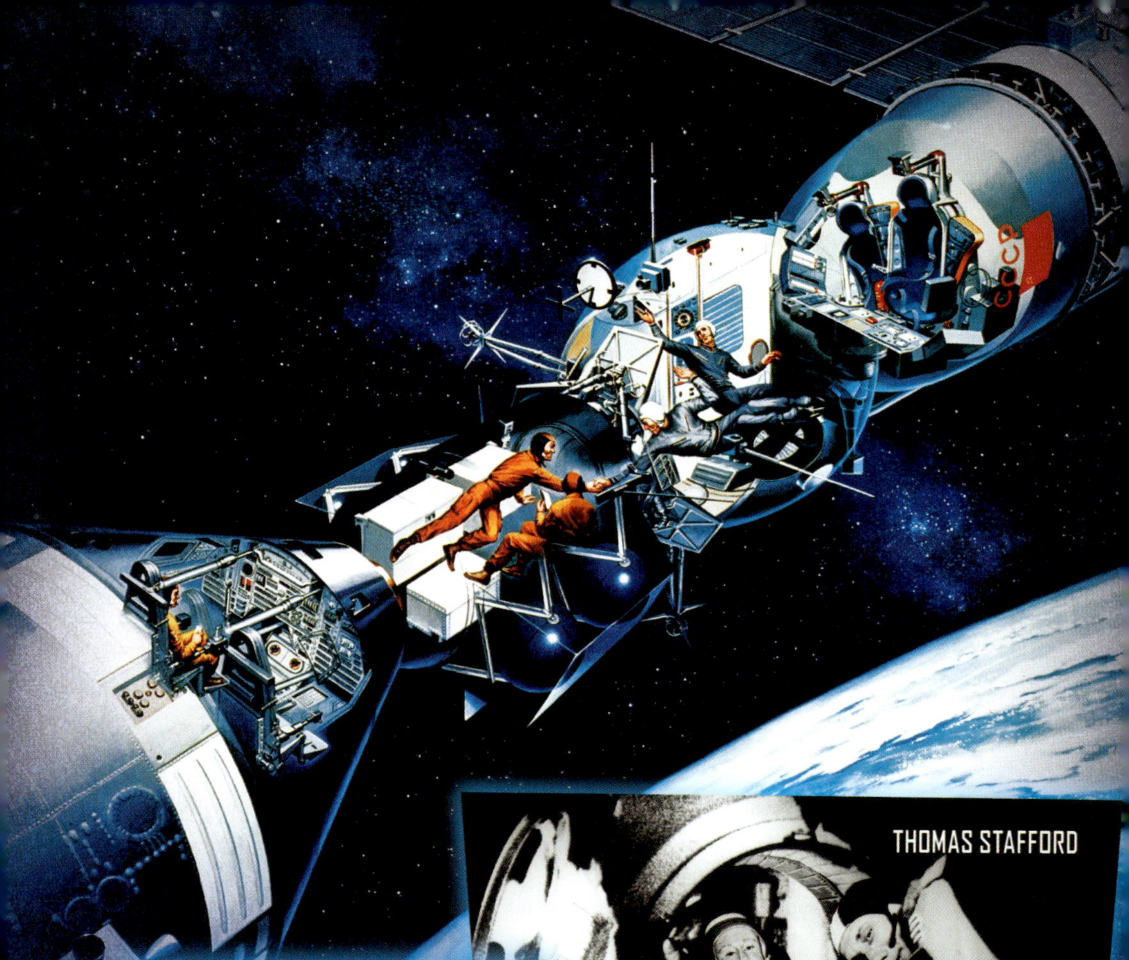

THE APOLLO-SOYUZ TEST PROJECT WAS THE FIRST MISSION TO HAVE DIFFERENT COUNTRIES WORKING TOGETHER IN THEIR OWN SPACECRAFT. THE U.S. AND SOVIET SPACECRAFT DOCKED ON JULY 17, 1975. THIS HISTORIC MISSION IS SEEN AS THE END OF THE SPACE RACE.

HANDSHAKE IN SPACE

During the Apollo-Soyuz Test Project, the United States launched an Apollo command module with Deke Slayton, Vance Brand, and Thomas Stafford aboard. The Soviets sent up a Soyuz rocket with Alexei Leonov and Valeri Kubasov. They trained together for years beforehand, learning each other's languages and becoming close friends. When the Apollo-Soyuz spacecraft docked and the hatches opened, Soviet commander Alexei Leonov and U.S. commander Thomas Stafford shook hands to mark the occasion.

THE FAB FIVE

The next step was a reusable rocket, so NASA came up with the space shuttle program. Shuttles were launched like a rocket and landed like an airplane on a runway. They could be used again, although they were limited to Earth's orbit.

NASA had five space shuttles: *Atlantis*, *Challenger*, *Columbia*, *Discovery*, and *Endeavor*. From 1981 to 2011, they flew 135 missions and helped build the International Space Station (ISS). Among their many tasks, the shuttles carried astronauts and equipment to and from the space station, repaired satellites, and performed research.

The program ended in 2011 because the remaining shuttles had 30 years of wear and tear and it was getting too difficult to maintain them. Also, the United States wanted newer shuttles that could move farther out into space.

INTERSTELLAR INFORMATION

THE **PROTOTYPE** SPACE SHUTTLE WAS NAMED *ENTERPRISE*, AFTER THE SHIP IN THE *STAR TREK* TELEVISION SHOWS AND MOVIES. IT'S NOW ON DISPLAY AT THE INTREPID MUSEUM IN NEW YORK CITY.

ATTACHED TO TWO ROCKETS AND A FUEL TANK, THE SPACE SHUTTLE *ENDEAVOR* IS READY FOR LIFTOFF AT THE KENNEDY SPACE CENTER IN FLORIDA IN 2007.

CHALLENGER ACCIDENT, 1986

TRAGEDY STRIKES

Not all of the shuttles survived. The *Challenger* exploded on January 28, 1986, just 73 seconds after liftoff. Unusually cold weather had damaged its rockets. All seven crew members were killed. Decades later, *Columbia* broke apart as it reentered Earth's atmosphere on February 1, 2003, killing all seven crew members. During the launch, a piece of foam hit the wing. This made a small hole that let hot gas into the shuttle, causing *Columbia* to burn up.

A GLOBAL EFFORT

The United States and the Soviet Union weren't the only countries interested in space. The European Space Agency (ESA), created in 1975, combined scientists and **technology** from 22 countries. The ESA has launched many probes to study Earth, including its environment, clouds, radiation (waves of unseen energy), polar ice, gravity, and weather.

The ESA has also studied the sun and other planets. Ulysses was launched in 1990 to chart the space above and below the sun's poles. The spacecraft SOHO went up to study the sun five years later. The Mars Express was launched in 2003. It was Europe's first Red Planet orbiter. Two years later, the Venus Express was Europe's first Venus orbiter. Cassini, launched in 1997, carried a surface probe called Huygens to Saturn's moon Titan.

INTERSTELLAR INFORMATION

THE ESA'S ROSETTA PROBE LAUNCHED IN 2004. TEN YEARS LATER, IT BECAME THE FIRST CRAFT TO SEND A LANDER TO THE SURFACE OF A COMET.

THIS IS A PHOTO OF THE SURFACE OF SATURN'S MOON TITAN, TAKEN BY THE CASSINI SPACE PROBE. AFTER OBSERVING SATURN AND ITS MOONS FOR MORE THAN 17 YEARS, CASSINI CRASHED IN 2017.

WORKING ON THEIR OWN

Created in 1993, the China National Space Administration oversees that country's space activities. Yang Liwei was the first Chinese national in space in 2003, and the country has done many manned launches since then. China landed a rover on the moon in 2013. It took pictures and studied soil. In 2019, China's Chang'e 4 became the first spacecraft to land on the side of the moon facing away from Earth. Unlike other space agencies, China doesn't work on the ISS.

GATEWAY TO THE STARS

The ISS was built between 1998 and 2011. It sits in low Earth orbit and is the largest structure humans have ever put into space. Many different scientists and astronauts have lived there since November 2, 2000, and they've conducted thousands of experiments.

The ISS was built with help from 15 countries, especially the United States, Russia (part of the former Soviet Union), the ESA, Canada, and Japan. The station is expected to continue working until at least 2024 and maybe as long as 2028. Depending on its condition, NASA may crash it into an ocean on Earth. If it's in good shape, the ISS could be pushed into a higher orbit around Earth, where it could survive for many more years.

INTERSTELLAR INFORMATION

AMERICAN ASTRONAUT SCOTT KELLY AND RUSSIAN COSMONAUT MIKHAIL KORNIENKO SHARE THE RECORD FOR MOST DAYS IN A ROW IN SPACE. THEY SPENT 340 DAYS IN SPACE DURING AN ISS MISSION FROM 2015 TO 2016.

THE ISS CIRCLES EARTH EVERY 90 MINUTES, FLYING AT A HEIGHT OF 248 MILES (399.1 KM) AND A SPEED OF 17,500 MILES (28,163.5 KM) PER HOUR.

LIVING IN SPACE

Generally, there are three to six people living on the space station, although it's held as many as 13 for short periods. Crew members must spend at least two hours each day focusing on exercise so that scientists can see how their bodies react to zero gravity. Astronauts have reported changes in their bones, muscles, and vision after spending months in space. More than 230 people from 19 different countries have visited the station.

CATCHING SOME RAYS

Ever since scientists realized that the sun is the center of our solar system, they've tried to learn as much as possible about it. It's a very hot ball of gasses, mostly hydrogen and some helium. The sun's core is about 27 million°F (15 million°C). It's 93 million miles (149.7 million km) away from Earth.

Many countries have sent probes to the sun. SOHO was a joint mission between NASA and the ESA launched in 1995. It studied the sun's structure and helped to predict solar outbursts. TRACE was launched in 1998 and provided information about the sun's corona, or outer layer. NASA's Genesis was launched in 2001 and was the first to capture a sample of solar wind, or tiny particles that come from the sun.

INTERSTELLAR INFORMATION

SCIENTISTS THINK THE SUN WILL KEEP BURNING FOR ANOTHER 6.5 BILLION YEARS BEFORE IT RUNS OUT OF ENERGY AND THEN GROWS MUCH DIMMER.

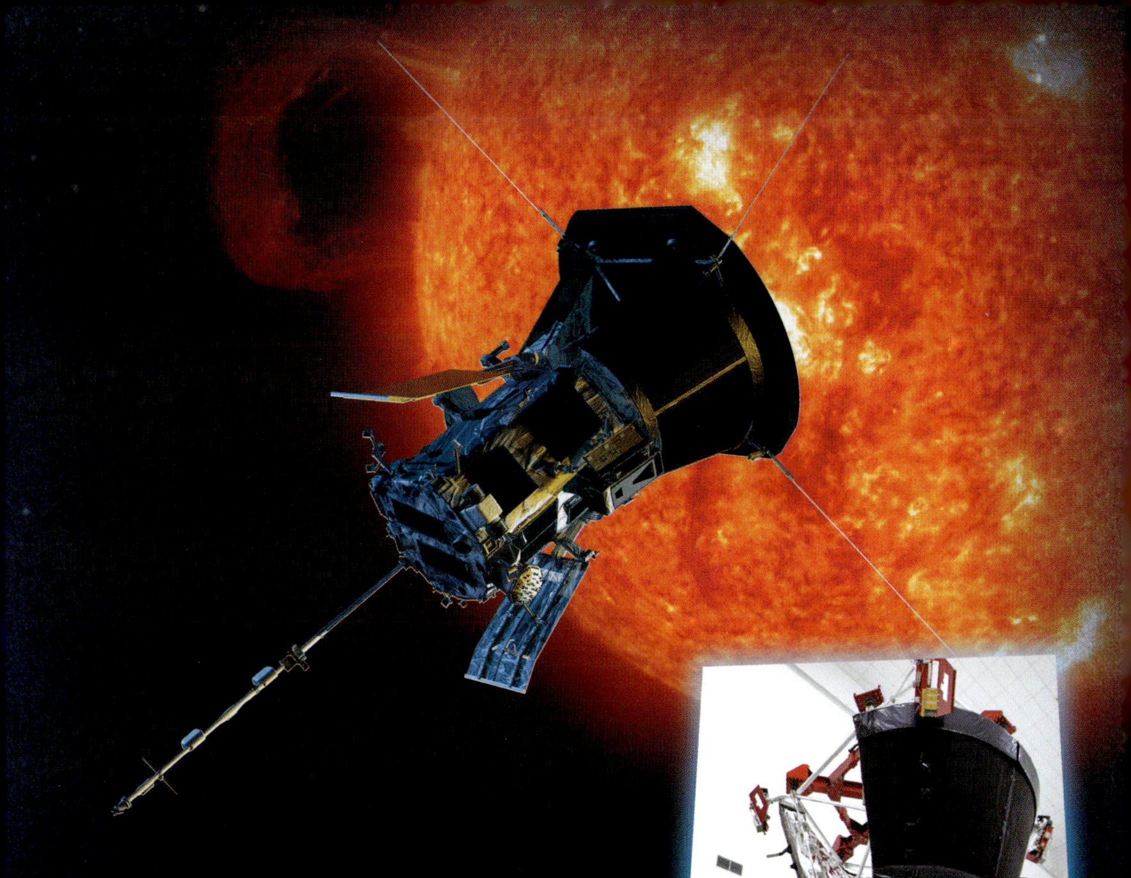

THE PARKER SOLAR PROBE, SHOWN AT RIGHT ON EARTH GETTING READY FOR LAUNCH, HAS SINCE FLOWN CLOSER TO THE SUN THAN ANY OTHER CRAFT IN HISTORY.

FLYING CLOSE TO THE SUN

NASA's Parker space probe came within 15 million miles (24 million km) of the sun in November 2018, flying at a record speed of 213,200 miles (343,100 km) per hour. It later flew through the sun's corona, thanks to a special foam shield called the thermal protection system that helps it handle temperatures of nearly 2,600°F (1,426°C). The Parker probe helps scientists better understand the sun's radiation, solar flares, and solar winds.

THE RED PLANET

Since Mars is the closest planet to Earth, scientists think that could be the next place to send astronauts. First, they want to learn all they can. Vikings 1 and 2 landed on Mars in 1976 and took photos and air and soil samples. However, it was the rovers, which could drive around the surface, that provided the most information, starting with Pathfinder and Sojourner in 1997. Spirit and Opportunity followed in 2004 and Curiosity in 2012. The Phoenix lander arrived in 2008 and explored the planet's north pole.

InSight, a robotic explorer, landed on Mars in 2018. It studies the planet's underground. In 2019, Curiosity found mud samples more than 3 billion years old. This clue could help scientists understand the ancient history of water, and maybe even life, on Mars.

INTERSTELLAR INFORMATION

OTHER COUNTRIES BESIDES THE UNITED STATES HAVE ALSO REACHED MARS. THE SOVIET UNION LANDED PROBES ON MARS IN THE 1970S. MORE RECENTLY, THE ESA AND INDIA HAVE ALSO REACHED MARS'S ORBIT.

THE CURIOSITY ROVER HAS A ROBOTIC ARM THAT ENABLES IT TO TAKE PHOTOS ... EVEN SELFIES! HERE IS CURIOSITY AT A MARTIAN DRILL SITE IN 2019.

WHAT'S THE WEATHER LIKE?

Mars has a thinner atmosphere than Earth that's made up mostly of carbon dioxide. There's a little bit of oxygen, but not nearly enough for humans to breathe. It's also colder than Earth because it's farther away from the sun, with an average temperature of -80°F (-62°C) although it can vary widely. Mars has heavy dust storms and radiation. However, recent probes have found underwater ice on the planet, meaning astronauts might have a source of water.

WHAT'S NEXT?

It's been 50 years and counting since anyone from Earth visited the moon—but that may be about to change. With Project Artemis, NASA is planning to land the first woman and the next man on the moon by 2024.

Using a powerful new rocket called the Space Launch System, NASA wants to build Gateway, a space station that will orbit the moon. There, astronauts could live and do experiments. Private companies such as SpaceX and Blue Origin are working with NASA to help create rockets and equipment to let astronauts travel back and forth from Gateway to the surface of the moon. From Gateway, NASA also hopes to be able to send humans to Mars. Will you be the first one to go to Mars? It's possible!

INTERSTELLAR INFORMATION

IN GREEK MYTHOLOGY, ARTEMIS WAS THE SISTER OF APOLLO. NOW, BOTH NASA PROJECTS NAMED AFTER THE TWO SIBLINGS WILL CARRY ASTRONAUTS TO THE MOON.

THE SPACEX FALCON HEAVY ROCKET LAUNCHED ON ITS FIRST TEST FLIGHT FROM PAD 39A AT KENNEDY SPACE CENTER—THE SAME PAD MOST OF THE APOLLO ROCKETS USED ON THEIR WAY TO THE MOON.

PRIVATE ROCKETS

The Falcon 9 and the Falcon Heavy, designed by SpaceX, are commercial rockets that can take supplies and satellites into space. The Falcon Heavy's boosters can detach when their fuel is spent. They fall back to Earth and can land safely so they can be used again. SpaceX is already working on a new, bigger rocket called the Big Falcon Rocket, which could someday carry passengers into space and maybe travel to Mars.

GLOSSARY

accurate: free from mistakes

aeronautics: a science that deals with airplanes and flying

galaxy: a large group of stars, planets, gas, and dust that form a unit within the universe

interstellar: having to do with the space between stars

lunar module: a space vehicle designed to carry astronauts from the main part of the ship to the surface of the moon and back

missile: a rocket used to strike something at a distance

Nazi: a member of a German political party that controlled Germany from 1933 to 1945 under Adolf Hitler

probe: an unmanned spacecraft that travels through space to collect science information

prototype: the first model on which later models are based

rover: a vehicle used to explore the surface of a planet or moon

technology: tools, machines, or ways to do things that use the latest discoveries to fix problems or meet needs

verge: a point where something is about to happen

FOR MORE INFORMATION

BOOKS

Fabiny, Sarah. *What Is NASA?* New York, NY: Penguin Workshop, 2019.

Hubbard, Ben. *Space Race: The Story of Space Exploration to the Moon and Beyond.* Hauppauge, NY: B.E.S. Publishing, 2019.

Solar System. New York, NY: DK Children, 2016.

WEBSITES

How Old Are Galaxies?
spaceplace.nasa.gov/galaxies-age/en/
This NASA SpacePlace site examines galaxies and what we know about them.

The Space Race
www.dkfindout.com/us/space/space-race/
Learn more about the Space Race and its projects on this website.

The Sun
spaceplace.nasa.gov/menu/sun/
Find out more about the sun on this NASA SpacePlace site.

Publisher's note to educators and parents: Our editors have carefully reviewed these websites to ensure that they are suitable for students. Many websites change frequently, however, and we cannot guarantee that a site's future contents will continue to meet our high standards of quality and educational value. Be advised that students should be closely supervised whenever they access the internet.

INDEX